The World with Zero Postulation

Subhajit Ganguly

V M A

The World with Zero Postulation

(Also known as: Zero Postulation and the Laws of Nature)

V M A Publications

Science Books

First published in 2014

Copyright © Subhajit Ganguly 2012

All rights reserved.

ISBN-13: 978-1499285505
ISBN-10: 1499285507

All rights reserved. No part of this book may be reproduced in any form or by any electronic or mechanical means, including information storage and retrieval systems, without written permission from the author, except in the case of a reviewer, who may quote brief passages embodied in critical articles or in a review. The information in this book is distributed on an "as is" basis, without warranty.

Zero postulation and the Laws of Nature
Subhajit Ganguly

DEDICATION

To the Mysteries of Nature…

ZERO POSTULATION AND THE LAWS OF NATURE
SUBHAJIT GANGULY

Contents

Acknowledgements ii

A Brief Overview of Theories 6

The Standard Model of Physics 57

A Few Open Problems Today 64

Deduction and Emergence 71

Zero Postulation and

The Theory of Abstraction 78

ZERO POSTULATION AND THE LAWS OF NATURE
SUBHAJIT GANGULY

ACKNOWLEDGMENTS

This book is a culmination of the good wishes of many individuals, without whom nothing would have been possible. I take this opportunity to thank them all from the bottom of my heart. Hope my endeavor does justice to all the good wishes and aspirations surrounding it.

1

A Brief Overview of Theories

In the modern world, the first scientific attempts to find order in chaos in the universe were probably made by Nicolaus Copernicus (1473–1543). In his treatise *De revolutionibus orbium coelestium* (*On the Revolutions of the Heavenly Spheres*) he established the heliocentric theory. He boldly went against the established norms of the Renaissance Europe that held the Earth as the centre of the universe. *De revolutionibus orbium coelestium,* which was published just before his death, put forward a seminal theory that dislodged the geocentric model of the universe and established a heliocentric view, with the use of mathematics.

Copernicus, in his book, Copernicus argued that the universe comprised eight spheres. The outermost consisted of motionless, fixed stars, with the Sun motionless at the center. The known planets revolved about the Sun, each in its own sphere, in the order: Mercury, Venus, Earth, Mars, Jupiter, and Saturn. The Moon, however, revolved in its sphere around the Earth. What appeared to be the daily revolution of the Sun and fixed stars around the Earth was actually the Earth's daily rotation on its own axis. He was not entirely able to free himself of ancient Ptolemaic system, however, and retained the idea that all the orbits of celestial bodies must be perfect circles. He also believed in the unobserved crystalline spheres. Despite this, his work, which put the Sun, rather than the Earth, at the centre of the known universe of the day, was a serious blow to Aristotle's science and thus helped in bringing about the Scientific Revolution.

Johannes Kepler (1571 –1630) took the Scientific Revolution forward. His first major scientific work was perhaps the ***Mysterium Cosmographicum*** (*The Cosmographic Mystery*). It was the first published defense of the Copernican system. Kepler's views of the known universe were influenced by his experiments with 3-dimensional polyhedral and Platonic solids. **Fig. 1** shows an example of a skeletal polyhedron. The edges of a simple polyhedron must meet the following criteria:

1. An edge joins just two vertices.
2. An edge joins just two faces.

In fact, Kepler is also credited to have contributed a great deal in the knowledge of what are known as the Kepler–Poinsot polyhedron **(Fig. 2)**. In simple geometry, a polyhedron (plural polyhedra or polyhedrons) is a solid in three dimensions with flat faces, straight edges and sharp corners or vertices.

A Platonic solid in Euclidean geometry is a regular, convex polyhedron with congruent faces of regular polygons and the same number of faces meeting at each vertex. Five solids meet those criteria, and each is named after its number of faces – tetrahedron (four faces), cube or hexahedron (six faces), octahedron (eight faces), dodecahedron (twelve faces) and icosahedrons (twenty faces). Kepler attempted to relate the five extraterrestrial planets known at that time to the five Platonic solids.

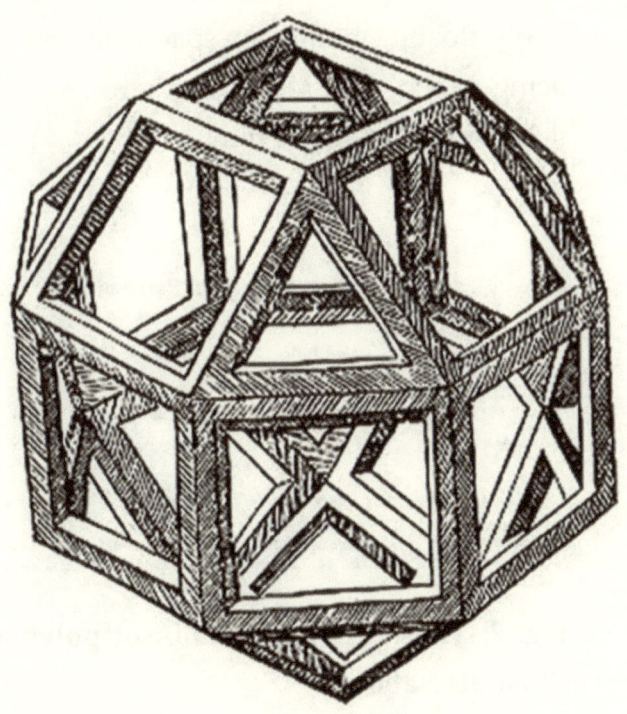

Fig. 1: The first ever printed version of the icosidodecahedron, by Leonardo da Vinci as apeared in the "Divina Proportione" by Luca Pacioli 1509.

It was in his work *Astronomia nova* that he proposed the elliptical path of the planets. He was the first to consider elliptical paths instead of circular ones. Kepler treated the planets as freely floating bodies in space, instead of them being attached to rotating spheres, as had been done by others before him.

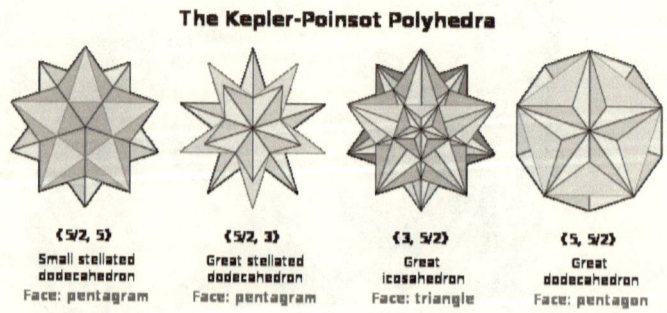

Fig. 2: The four Kepler–Poinsot polyhedra are illustrated above.

Kepler's *Astronomia nova* is regarded as one of the most important works of the Scientific Revolution of the Middle Ages. He discarded the Aristotelian notion that heavy bodies moved naturally towards the Earth, it being the centre of the universe. He used gravity instead (though his notion of gravity was that it was an attractive force similar to magnetism) to explain the movement of bodies towards the Earth. He states:

"Gravity is a mutual corporeal disposition among kindred bodies to unite or join together; thus the earth attracts a stone much more than the stone seeks the earth. (The magnetic faculty is another example of this sort).... If two stones were set near one another in some place in the world outside the sphere of influence of a third kindred body, these stones, like two magnetic bodies, would come together in an intermediate place, each approaching the other by a space proportional to the bulk [*moles*] of the other.... For it follows that if the earth's power of attraction will be much more likely to extend to the moon and far beyond, and accordingly, that nothing that consists to any extent whatever of terrestrial material, carried up on high, ever escapes the grasp of this mighty power of attraction."

Kepler also believed that the attractive force between two bodies would be mutual and proportional to the bulk of the bodies. He, however, did not say anything about how the gravitational force varied with distance. He is best known today for his **laws of planetary motion** (which also provided one of the concepts for Isaac Newton's theory of **Universal Gravitation**), based on his works *Astronomia nova* and *Harmonices Mundi* (*Harmonies of the world*). Kepler's laws of planetary motion describe the motion of the planets round the Sun. The modern form of the laws is:

1. The orbit of a planet is an ellipse with the Sun at one of the two foci.
2. A line segment joining a planet and the Sun sweeps out equal areas during equal intervals of time.

3. The square of the orbital period of a planet is proportional to the cube of the semi-major axis of its orbit.

Kepler investigated the motion of the planet Mars in order to arrive at the concept of elliptical path of the planets.

Galileo Galilei (1564–1642) was a contemporary of Johannes Kepler, who played a major role in the Scientific Revolution and the formation of later theories. Albert Einstein said about him 'Galileo ... is the father of modern physics—indeed of modern science'. Galileo's single most important contribution to the theorizing process is perhaps that he boldly applied mathematics to it. Galileo was one of the first modern thinkers to clearly state that the laws of nature are mathematical. In *The Assayer* he wrote 'Philosophy is written in this grand book, the universe ... It is written in the language of mathematics, and its characters are triangles, circles, and other geometric figures…'

He was ahead of his contemporaries in combining mathematics and experiments into the formation of theories. Scientific works of his time mostly consisted of qualitative studies and he ushered in a new era by implementing modern scientific methods in theoretical physics. He looked at the sky through his telescope and viewed a hitherto unseen world. His ideas in astronomy were vastly ahead of his time and that drew the wrath of the church and other established authorities.

However, nobody could subdue the indomitable spirit of the scientist in him and he went on to conclude his life's work into the *Discorsi e dimostrazioni matematiche, intorno à due nuove scienze* (Discourses and Mathematical Demonstrations Relating to Two New Sciences). In the *Two New Sciences*, he discussed kinematics (or the movement of bodies) and the strength of materials. The book is written as a series of discussions, over a span of four days, among two philosophers and a layman. Salviati holds the Copernican position and voices some of Galileo's own direct views, Sagredo is a layman (intelligent and initially neutral) and Simplicio (a contemporary conservative philosopher). The work spanned most of contemporary science.

In *Two New Sciences*, Galileo alters the fundamental nature of observing the rate of falling bodies by using an inclined ramp rather than free fall. He did away with Aristotelian notion that heavier bodies fell quicker than lighter ones. Galileo was the first to express clearly that a constant acceleration (acceleration due to gravity) was involved in a freefall. He was able to measure this acceleration accurately by slowing it down using an inclined plane.

According to Aristotle, an arrow shot straight up in the air came back to the spot from which it was shot. This, according to him, was ample proof that the Earth was immovable. Galileo refuted this by stating the example of a ball thrown vertically in the air or an arrow shot vertically from the deck of a moving ship, both returns to the same spot. Some of Galileo's most influential experiments regarding falling bodies were those describing the relativity of motions. 'One motion may be superimposed upon another without effect upon either...' (under the right conditions) is an idea that was proposed by him regarding this. This idea would later become the basis for *Newton's first law of motion* or the *law of inertia*.

Galileo was also the first in describing the *square-cube law* (or *cube-square law*) states that, as a shape grows in size, its volume grows faster than its area. This is a very important concept, pertaining to the real world, which has implications in varied fields that include mechanical engineering, biology, etc. This law is the reason why it is increasingly difficult for larger mammals (like the elephant) to cool themselves and for engineers and designers to build taller structures (like the skyscrapers). Among other things, Galileo's ***Two New Sciences*** also contains a discussion of infinity.

Isaac Newton (1642 – 20 March 1727) was born within a couple of months of Galileo Galilei's death. Regarded as one of the greatest scientists of all time, he took giant strides for furthering the Scientific Revolution. He was among the first to use modern methods of scientific inquiry and can, therefore, be truly held as the bridge between the science of the Middle Ages and that of the modern day. However, Newton himself had been rather modest of his own achievements, famously writing in a letter to Robert Hooke in February 1676 'If I have seen further it is by standing on the shoulders of giants.'

Newton's contribution in the fields of physics and mathematics is extraordinary. He was the one, who laid the foundations for classical mechanics and optics. He also developed (along with his peers, particularly Gottfried Leibniz) the mathematics to describe his theories. The modern form of infinitesimal calculus is a case in point. Infinitesimals have been used to express the idea of objects so small that there is no way to see them or to measure them. In simple words, an infinitesimal quantity is one, which is smaller than the smallest that you can think of. He also made notable contributions to the generalized binomial theorem among other things.

He made systematic study of light and its properties. The dispersion of light and the recombination of the seven colours of the visible spectrum were discovered by him. Newton proposed a corpuscular theory of light in his *Optiks* (1704). His corpuscular theory is however vastly different from the modern particle concept of light or the photon and they are not to be confused as the same.

The *Optiks* was a remarkable leap even from the point of view of evolution of the theorizing process in modern science. It moved beyond forming hypotheses and relied heavily on experimental observations. The results in the book were not deduced using deduction from previously discovered theorems; it defined new axioms to describe properties of matter and energy. These new axioms were the results derived from scientifically devised experiments. Newton is undoubtedly one of the greatest experimenters of all time. His works proved to be extremely significant in establishing the modern experiments-based physics. No longer were mere hypotheses enough 'proof' of theories, if experimental results showed otherwise, any hypothesis was to be rejected.

Zero Postulation and the Laws of Nature
Subhajit Ganguly

Newton contributed a great deal in placing this idea on solid grounds.

Isaac Newton's greatest work is, however, the ***Philosophiæ Naturalis Principia Mathematica*** (Mathematical Principles of Natural Philosophy), first published in 1687. It is in this book that we find the famous Newton's laws of motion and his law of universal gravitation. The ***Principia*** laid the foundation of classical mechanics and that of modern physics. The book is divided into three parts. The first part deals with the motion of bodies, the second part a continuation of the work laid down in the first part, and part three deals with gravitation. Parts one and three have proved to be more successful in passing the tests of time than part two. Part two was mainly written to refute the Cartesian theory of vortices by René Descartes, which held that planetary motions were produced by the whirling of fluid vortices that filled interplanetary space and carried the planets along with them.

Part three of the Principia is also important from the point of view of it containing **the Rules of**

Reasoning in Philosophy. These rules are very important in understanding the evolution of the theorizing process in science. The rules are:

Rule 1: *We are to admit no more causes of natural things than such as are both true and sufficient to explain their appearances.*

Rule 2: *Therefore to the same natural effects we must, as far as possible, assign the same causes.*

Rule 3: *The qualities of bodies, which admit neither intensification nor remission of degrees, and which are found to belong to all bodies within the reach of our experiments, are to be esteemed the universal qualities of all bodies whatsoever.*

Rule 4: *In experimental philosophy we are to look upon propositions inferred by general induction from phenomena as accurately or very nearly true, not withstanding any contrary hypothesis that may be imagined, till such time as other phenomena occur, by which they may either be made more accurate, or liable to exceptions.*

Newton also listed a number of mostly astronomical data and named them the

Phenomena. It was on these *Phenomena* that he depended in order to draw inferences later in the book. The third part thus laid down the basic principle or method that is used to form theories, by clearly defining the roles of scientific reasoning and experimentation in the process.

The law of universal gravitation states that any two bodies in the universe attract each other with a force that is directly proportional to the product of their masses and inversely proportional to the square of the distance between them. This was the genius of Newton in seeing the same underlying force that causes all bodies in the universe, big or small, to attract each other. Using his new theory, he was able to predict with considerable accuracy the motion of planets, as well as the tiniest objects known to man in those days.

He himself was not fully comfortable, however, with the force of gravity. In a letter, he wrote 'That one body may act upon another at a distance through a vacuum without the mediation of anything else, by and through which their action and force may be conveyed from one another, is to me so great an absurdity that, I believe, no man who has in philosophic matters a competent faculty of thinking could ever fall into it.' He was never able to assign a cause behind the existence of gravity. In the second edition of the *Principia* he wrote 'I have not yet been able to discover the cause of these properties of gravity from phenomena and I feign no hypotheses... It is enough that gravity does really exist and acts according to the laws I have explained, and that it abundantly serves to account for all the motions of celestial bodies.'

It is however to be noted that the idea of gravity was not an entirely new concept outside Europe. In **Siddhantha Siromani** (*Supreme Results*) gravity was described by the 11th century Indian Mathematician Bhaskaracharya in the following terms:

Aakrishti sakthischa mahee thayaa yathkhastham guru swa abhimukham swa sakthyaa . aakrushyathe thath pathathi iti bhaathi same samanthaath kwa pathathi ayam khe

It says that the earth attracts the objects in the sky by its own force towards itself. He discusses the forces between the celestial bodies using a question: Where can the celestial bodies fall since they attract each other?

It was much after the death of Isaac Newton that modern science began to learn about electricity and magnetism in details. Around the beginning of the nineteenth century a vague relationship between electricity and magnetism began to be understood, especially due to the experiments conducted by Hans Christian Ørsted, André-Marie Ampère, Jean-Baptiste Biot, Johann Carl Friedrich Gauss, Félix Savart, Michael Faraday, etc. However, the knowledge that was being gathered was vastly chaotic. James Clerk Maxwell (1831–1879) organized the knowledge by formulating a few general principles combining, both electricity and magnetism. His work was about finding the underlying principles that causes the various electrical and magnetic phenomena. Maxwell helped develop the Maxwell–Boltzmann distribution, which is a statistical means of describing aspects of the kinetic theory of gases.

Between 1861 and 1862, Maxwell published his famous four-part paper *On Physical Lines of Force*. Maxwell combined the ideas of electricity and magnetism and derived the equations of electromagnetism. Electromagnetism was a completely new concept that he successfully described and his work is a prime example of seeing the underlying principle behind two seemingly dissimilar ideas. While a relation between electricity and magnetism was emerging, he was the first to describe this relation in concrete and *simple* terms.

On Physical Lines of Force is considered to be one of the most important works in physics (and science in general). It has influenced later researchers immensely and ushered in a new era of classical electrodynamics. Vector calculus is one area in mathematics that this work made notable contributions too.

In 1865, Maxwell published *A Dynamical Theory of the Electromagnetic Field*. He used concept of displacement current, which he had introduced in his earlier paper *On Physical Lines of Force* and derived the electromagnetic wave equation, which describes the propagation of electromagnetic waves through a medium or through vacuum. He also established a relation between the speed of light inside a medium (or vacuum) and electromagnetic properties of the medium. Maxwell's equations describe how electric and magnetic fields are generated and altered by each other and by charges and currents.

In *Part VI* of his 1864 paper titled *Electromagnetic Theory of Light*, he combined displacement current with some of the other equations of electromagnetism and obtained a wave equation with a speed equal to the speed of light. He commented 'The agreement of the results seems to show that light and magnetism are affections of the same substance, and that light is an electromagnetic disturbance propagated through the field according to electromagnetic laws.'

Maxwell calculated that the speed of propagation of an electromagnetic field is approximately equal to the speed of light. He commented 'We can scarcely avoid the conclusion that light consists in the transverse undulations of the same medium which is the cause of electric and magnetic phenomena.' This concept of the speed of light being equivalent to the speed of propagation of an electromagnetic field is a very important concept that laid the foundation of many important later works, including Albert Einstein's Theory of Special Relativity. On Maxwell's work, Einstein would say '…it took physicists some decades to grasp the full significance of Maxwell's discovery, so bold was the leap that his genius forced upon the conceptions of his fellow-workers.' To summarize James Clerk Maxwell's contribution on the evolution of the idea of a finite and constant speed of light, he said that light was an electromagnetic wave, and therefore travelled at the speed c appearing in his theory of electromagnetism.

Maxwell believed that the propagation of light required a medium for the waves, dubbed the *luminiferous aether* (light-bearing aether*)*, which would always be stationary. Aether was something that would permeate all space and yet be apparently undetectable by mechanical means. The existence of such a medium proved impossible to reconcile with experiments such as the Michelson–Morley experiment (**Fig. 3**), later on. Finally, Albert Einstein (1879–1955), with his Special Theory of Relativity, managed to do away with the requirement of such a medium that would be necessary to make Maxwell's equations valid. Aether, the postulated medium for the propagation of light was no longer necessary.

Note that Einstein sometimes used the word *aether* for the gravitational field within general relativity, which he formulated later. However, the term *aether* never gained widespread support ever again. About aether he said in ***Ether and the Theory of Relativity*** in 1920:

We may say that according to the general theory of relativity space is endowed with physical qualities; in this sense, therefore, there exists an aether. According to the general theory of relativity space without aether is unthinkable; for in such space there not only would be no propagation of light, but also no possibility of existence for standards of space and time (measuring-rods and clocks), nor therefore any space-time intervals in the physical sense. But this aether may not be thought of as endowed with the quality characteristic of ponderable media, as consisting of parts which may be tracked through time. The idea of motion may not be applied to it.

In 1905, Einstein published his work on the nature of light itself. The photoelectric effect is the observation that many metals emit electrons when light shines upon them. The velocity of the emitted electrons does not depend upon the intensity of light falling on the surface of the metal, but on its frequency. This does not fit the classical wave model of light. In order to explain the *photoelectric effect* he took the help of the concept of the photon.

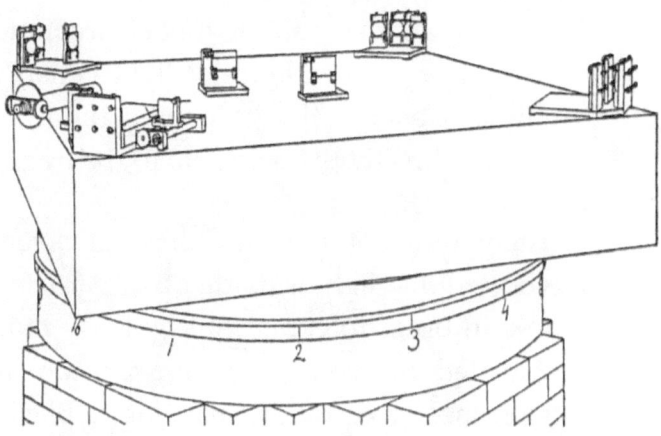

Figure 3: Michelson and Morley's experimental setup, mounted on a stone slab and floating in a pool of mercury.

A photon is an elemental particle (or quantum) that constitutes all form of electromagnetic radiation, including light. This new concept was similar to the corpuscular theory of light by Isaac Newton in the sense that both considered light to be made up of particles. However, the similarity ends there. The two concepts are different in all other respects. The modern photon concept was developed gradually by Einstein to explain experimental observations that did not fit the classical wave model of light. In particular, the photon model accounted for the frequency dependence of light's energy. The amount of energy that is carried by any given photon is directly proportional to the frequency.

Einstein was not the only one, however, who advocated that light is made up of particles. Max Plank, in order to explain Black-body radiation, formulated the concept of the photon, mostly out of sheer frustration than logic. Black-body radiation is the type of electromagnetic radiation within or surrounding a body in thermodynamic equilibrium with its environment, or emitted by a black body (an opaque and non-reflective body) held at constant, uniform temperature. The radiation has a specific spectrum and intensity that depends only on the temperature of the body. The Munich physics professor Philipp von Jolly advised Planck against going into physics, saying, 'In this field, almost everything is already discovered, and all that remains is to fill a few holes.' Plank, however, wanted to understand and learn about the mysteries of the world and did not (fortunately!) follow the professor's advice. He went on to become one of the key figures, who originated an entirely new branch of physics, namely the quantum theory.

Coming back to Einstein, he took the concept of the photon further and went on to show that any given photon has momentum. In his 1909 paper ***über die Entwicklung unserer Anschauungen über das Wesen und die Konstitution der Strahlung*** (*The Development of our Views on the Composition and Essence of Radiation*), he noted:

…Today, however, we regard the ether hypothesis as obsolete. A large body of facts shows undeniably that light has certain fundamental properties that are better explained by Newton's emission theory of light than by the oscillation theory. For this reason, I believe that the next phase in the development of theoretical physics will bring us a theory of light that can be considered a fusion of the oscillation and emission theories. The purpose of the following remarks is to justify this belief and to show that a profound change in our views on the composition and essence of light is imperative.

Albert Einstein was thus also the pioneer of the concept of wave–particle duality, in quantum mechanics in a way.

In 1905, Einstein published ***Zur Elektrodynamik bewegter Körper*** (*On the Electrodynamics of Moving Bodies*). It studied what we know as the Special Theory of Relativity. In it, he reconciled Maxwell's theory of electromagnetic radiation with the laws of mechanics. Einstein depended on two basic postulates in order to formulate this theory:

1. The laws of physics are invariant for any non-accelerating frame of reference (called an inertial frame of reference).
2. The speed of light has the same value in all inertial frames of reference, independent of the state of motion of the emitting body.

The second postulate, in particular, seems to be in direct contradiction with Newtonian classical mechanics. In that, it rewrote physics and took our understanding of the universe forward. The first postulate is an echo of Galileo's idea that the laws of nature should be the same for all observers that moves with constant speed relative to each other.

The implications of the Special Theory of Relativity seemed to defy *commonsense*. Newtonian theories assumed time to be a continuum that is absolutely constant for every part of the universe. Einstein's new theory was a severe blow to this concept. It propounded the idea of *time dilation*, which is an actual difference of elapsed time between two events, as measured by observers moving relative to each other (or differently situated from gravitational masses, as shown by the General Theory of Relativity). This was a paradigm shift in looking at the universe and it changed the course of physics forever. Newtonian mechanics was no longer sufficient to describe the universe.

Another *commonsense-defying* result that follows directly from the idea of *time dilation* is *length contraction*, which is the phenomenon of a decrease in length measured by an observer of objects which are traveling at any non-zero velocity relative to the observer. However, this effect is negligible at speeds very much smaller than the speed of light (speed of light in vacuum is 3×10^8 meters/second – 3 multiplied by 10 followed by 8 zeros!). At speeds comparable to the speed of light, *length contraction* can no longer be neglected.

In 1905 itself, Einstein published another paper ***Ist die Trägheit eines Körpers von seinem Energieinhalt abhängig?*** (*Does the Inertia of a Body Depend Upon Its Energy Content?*). In it, he put forward what is known as the most famous equation in physics, relating the mass of a body and the energy contained by it.

The Special Theory of Relativity is *special* because it is valid only in special cases, where inertial frames of reference are involved. It gives a good description of the world only when uniform velocities (or no acceleration) are involved. In order to incorporate his relativistic principles to describe the world in general, Einstein had to develop the General Theory of Relativity.

Zero Postulation and the Laws of Nature
Subhajit Ganguly

We can trace the beginning of the formation of this new theory to what Einstein himself described as *the happiest thought of my life*. In 1907, he struck upon this idea that a person would be unable to feel any gravitational field, when he/she is having a freefall. This *Equivalence Principle* may be regarded as the starting point of general relativity. This principle equates free fall with inertial motion, a vast shift from Newtonian principles. With the *Equivalence Principle* in his hand, Einstein could apply the rules of special relativity for a freefalling observer. He also introduced the concept of *gravitational time dilation*. Gravitational time dilation is the difference of elapsed time between two events as measured by observers differently situated from gravitational masses, in regions of different gravitational potential. The closer a clock is to the source of gravity (i.e., the lower the gravitational potential), lower is the rate at which time passes. This means that if we orbit the earth, we will be affected by this time dilation (and will age slightly less slowly, for speeds human beings have managed to travel at till now).

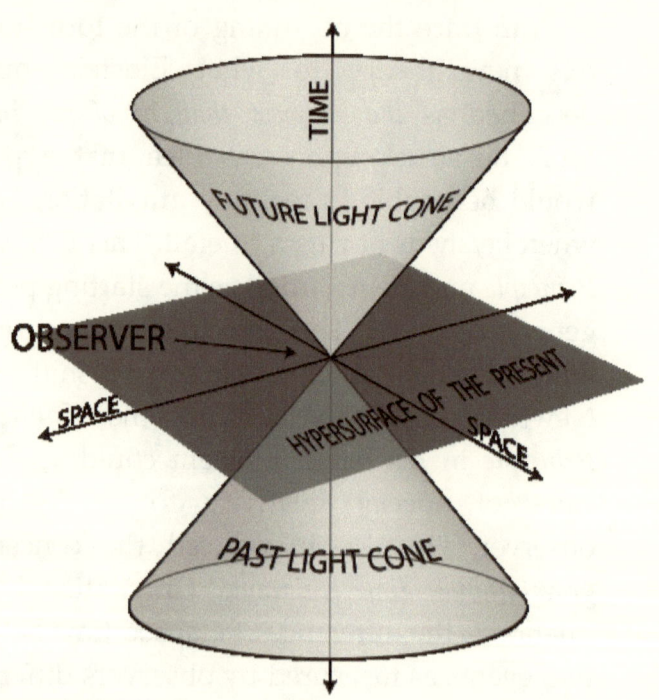

Fig. 4: The *world line* of an object is the unique path of that object as it travels through 4-dimensional *spacetime*. The concept of world line is distinguished from the concept of orbit or trajectory (such as the orbit of a planet in space or the trajectory of a car on a road map). The idea of world lines was pioneered by Hermann Minkowski.

Reaching the final form of the theory was delayed due to a non resolution of the *hole argument*, which assumed a meaning to separate spacetime coordinates. In other words, it assumed that spacetime could exist even without having any matter in it. Einstein was finally able to resolve this by doing away with the assumption. He later stated:

People before me believed that if all the matter in the universe were removed, only space and time would exist. My theory proves that space and time would disappear along with matter.

This idea was another revolution in our way of looking at the universe.

By 1915, Einstein had published the final form of the General Theory of Relativity. The view that this theory gives us is that sources of gravity (read bodies with mass) twist the fabric of spacetime (space and time can no longer be held as separate entities). In this spacetime, a freely moving or falling body always moves along a geodesic (the equivalent of a straight-line on a curved surface). Gravity, according to this theory, is not a force, but the result of the curvature of spacetime by a body, which has mass. The Sun, being a massive body, distorts the fabric of spacetime around itself. The Earth (and all the other planets) move according to these distortions and gives us the feeling that they are moving under gravity. Thus, gravity was reduced from being a force and described to be a result of the geometry of spacetime itself. The theory also predicted that due to the curvature of spacetime around a massive object, light too would bend around it, giving rise to *gravitational lensing*.

Einstein's next great radical change in the world of physics is the formulation of the *Bose-Einstein Statistics* that proved to be the *missing link* between the old quantum theory (as started by Max Plank) and its modern form (as started by Erwin Schrödinger later). Satyendra Nath Bose, who had earlier translated Einstein's papers on relativity from the original German to English, sent Einstein his own work in 1924. ***Planck's Law and the Hypothesis of Light Quanta***, as the work had been named by Bose, considered light to be a gas of photons that were indistinguishable from each other. In doing that, he was able to do away with the Maxwell–Boltzmann distribution, which was proving to be increasingly inaccurate in predicting the microscopic world. Einstein adopted the idea and extended it to atoms. This led to the prediction of the existence of phenomena which became known as *Bose–Einstein condensate* (*BEC*). The *Bose–Einstein condensate is a* dense collection of bosons (which are particles with integer spin, named after Bose). This state of matter of a dilute gas of bosons can be formed, when the gas is cooled to temperatures very close to absolute zero (that is, very near 0 K or −273.15 °C.

It is generally held that Einstein had his major scientific breakthroughs behind him by the 1930s. Although he was one of the founding fathers of the quantum theory, he was very apprehensive of the probabilistic approach that it had been taking lately. He liked the idea of a universe that can be predicted with certainty over the idea of a collection of chances that the quantum theory presented. He tried his best to disprove the quantum theory, but the mainstream physicists decided to ignore him in this.

His later years were spent in trying to prove the quantum theory wrong and to form a *theory of everything*. He called this theory that would be able to describe correctly all phenomena in the world as the *unified field theory*. Within the geometric formulation of general relativity, he attempted to incorporate electromagnetic phenomena. However, all his attempts to unify gravitation and electromagnetism proved to be a failure. The world of physics had moved on and introduced two new types of fundamental forces of nature, namely the strong and weak nuclear forces. Einstein did not assimilate these two forces in his attempted *unified field theory*. As it did not take into account quantum principles, his attempted *unified field theory* is also called the *classical unified field theory*. Many scientists have tried to arrive at a unified theory that can be built from classical physics. However, none of such attempts have met with success.

Among the notable physicists, who put the quantum theory on a strong footing was Erwin

Schrödinger (1887–1961). He helped a great deal in developing the *wave mechanics* and discovered the *wave equation* (also known as the Schrödinger equation). The paper in which he published the *wave equation* was **Quantisierung als Eigenwertproblem** (*Quantization as an Eigenvalue Problem*). This equation is regarded as one of the most celebrated equations in physics and it ushered in a new phase in the development of the quantum theory. Generally speaking, the *Schrödinger equation* is a partial differential equation that describes how the quantum state of a physical system changes with time. As *Newton's second law of motion* describes the movement of a macroscopic body, so does the *Schrödinger equation* at the subatomic level. It directly predicted quantization of energy and angular momentum of a system. This is another revolutionary aspect of it. Like Einstein, Schrödinger too was a reluctant proponent of the quantum theory. About the probability interpretation of quantum mechanics he said:

I don't like it, and I'm sorry I ever had anything to do with it.

Paul Dirac (1902–1984) was another physicist, who made great contributions towards the development of the quantum theory and the *quantum electrodynamics (QED)*. *QED* was the first theory that described how matter and light interacts. It is the *relativistic quantum field theory* of electrodynamics and is the quantum equivalent of classical electromagnetism. In order to describe all interactions, where electrically charged particles are involved, *QED* considers the transport of photons between the systems involved. Dirac was successful in marrying special relativity with the quantum theory fully for the first time as he formulated the *Dirac equation*, which describes the behaviour of fermions. Fermions are particles, which follow the *Fermi–Dirac statistics* and also following Pauli's exclusion principle (which again states that two identical fermions, or particles with half-integer spin, cannot occupy the same quantum state simultaneously). The *Dirac equation* also predicted the existence of antimatter. Generally speaking, antimatter is material composed of antiparticles, which have the same mass as particles of ordinary matter but have opposite charge. Encounters between particles and antiparticles lead to the annihilation of both, with their masses getting converted to energy.

All magnetic substances that we know consist of two poles, one south and one north. However, Dirac showed in his work that if a magnetic monopole (either north or south) exists in the universe, all electric charge in the universe will be quantized. What we find is that electric charge is indeed quantized. Magnetic monopoles have never been found though, till date.

Dirac's work on the quantization of gravity went on to form the basis of the *gauge theories* and *superstring theories* later on.

Werner Karl Heisenberg (1901–1976) further took the quantum theory forward by introducing the *uncertainty principle* named after him. Heisenberg's *uncertainty principle* puts a fundamental limit to the precision with which certain pairs of physical properties of a particle (known as complementary variables, such as position and momentum) can be known simultaneously. This uncertainty is not due to the precision of the instruments used but is a fundamental property of quantum systems. For example, if we are to know the value for the position of a particle with absolute certainty, its momentum can only be predicted with zero certainty. This uncertainty arises as, according to the quantum theory, it is impossible to measure anything without disturbing it. For instance, any attempt to measure a particle's position must randomly change its speed. The *uncertainty principle* is one of the fundamental properties of quantum systems.

Over the centuries, we have had only a handful of physicists, who excelled both theoretically and experimentally. Enrico Fermi (1901–1954) is one such rare example. He Fermi held several patents related to the use of nuclear power. In theoretical physics, he was one of the physicists, who formulated the *Fermi–Dirac statistics*, which we mentioned earlier. Fermi was instrumental in the postulation of a new type of particle, called the *neutrino*, along with Wolfgang Ernst Pauli (1900–1958). A *neutrino* (which means, the *little neutral one* in Italian) is an electrically neutral, elementary subatomic particle, having half-integer spin. It is a fermion that follows the *Fermi–Dirac statistics*. Fermi was also instrumental in formulating the *weak interaction*, which is the mechanism responsible for the *weak force*. The *weak force* is one of the four fundamental forces of nature, alongside the *strong force*, the *electromagnetic force*, and *gravity*. The *weak interaction* is responsible for both the radioactive decay and nuclear fusion of subatomic particles.

Niels Bohr (1885–1962) was another major pioneer of the quantum theory and his work shed new light on the structure of the atom. Before Niels Bohr, it was believed that inside the atom electrons revolve round the nucleus in orbits, as planets round the sun. However, such a model would make the atom unstable, as the electrons, being charged particles, would lose energy and spiral into the nucleus. Bohr did away with the possibility of this lack of stability of the atom by postulating that the electrons can occupy only definite energy levels. He was involved in a series of debates with Albert Einstein regarding the validity of the quantum theory.

Zero Postulation and the Laws of Nature
Subhajit Ganguly

With the development of the quantum theory, it was becoming excessively clear that dynamics at the subatomic level, at least, cannot be expressed as a single, unique trajectory or path. This was a fundamental difficulty that the quantum theory had to face for some time. In order to express quantum amplitude over infinitely many possible trajectories, Richard Phillips Feynman (1918–1988) formulated the *path integral*. The idea of *path integral* has proved crucial to the advancement of theoretical physics in a great many ways. Being symmetric between space and time, it lets an investigator to easily change coordinates between very different canonical descriptions of the same quantum system.

Feynman also made major contributions to the development of *quantum electrodynamics* (*QED*). The key components of Feynman's presentation of QED are three basic actions.

1. A photon goes from one place and time to another place and time.
2. An electron goes from one place and time to another place and time.

3. An electron emits or absorbs a photon at a certain place and time.

The *Feynman diagram* (or the pictorial representation of the mathematical expressions governing these transformations) is shown in **Fig. 5**.

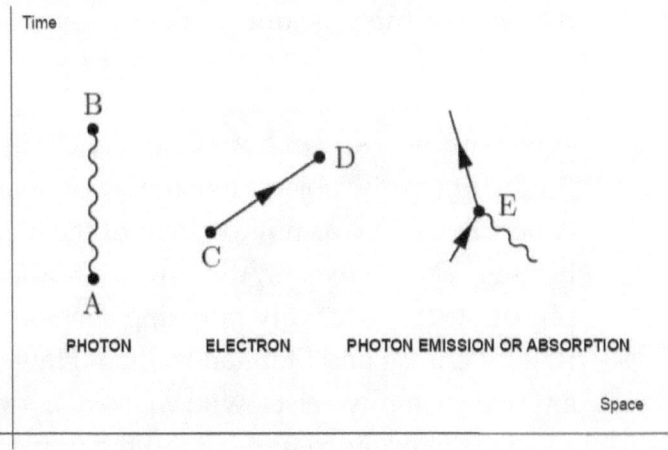

Fig. 5: Feynman diagram for transport of a photon, an electron, and for the emission or absorption of a photon, over space and time.

The *Feynman diagrams* in themselves are a very revolutionary concept in theoretical physics.

Richard Phillips Feynman also made significant contributions towards the fundamental foundations of concepts like quantum computing (where principles of the quantum world are involved directly, instead of the transistors that are used in ordinary computers) and nanotechnology (the engineering of functional systems at the molecular scale). One nanometer (nm) is one billionth, or 10^{-9}, of a meter.

While his own research was the pioneer in predicting exotic objects like the black hole, Albert Einstein was not too fond of the idea himself. That, however, did not deter other physicists from actively pursuing such objects. Roger Penrose and Stephen William Hawking are two such physicists, who worked along these concepts in order to understand the universe better. Both of these scientists have made significant contributions to general relativity and cosmology. Cosmology is the study of the universe at large and is concerned with its dynamics, its formation, its evolution, and its ultimate fate. Penrose and Hawking collaborated to work on what is now known as the *Penrose–*

Hawking gravitational singularity theorems that study the *singularities*, arising out of general relativity.

Fig. 6: A Penrose tiling, named after Roger Penrose who investigated these sets of non-periodic tilings in the 1970s.

ZERO POSTULATION AND THE LAWS OF NATURE
SUBHAJIT GANGULY

A singularity in solutions of the Einstein field equations is one of two things:

1. a situation where matter is forced to be compressed to a point (a space-like singularity)
2. a situation where certain light rays come from a region with infinite curvature (time-like singularity)

In general relativity, a *singularity* is a place that objects or light rays can reach in a finite time where the curvature becomes infinite, or space-time stops being a manifold.

While radioactive decay and nuclear fusion of subatomic particles was explained using the *weak force*, the *strong force* was brought into play in order to describe the stability of ordinary matter. The *strong force* accounts for the stability of elementary particles of ordinary matter, like the neutron and the proton as well as the stability of the atomic nucleus by binding neutrons and protons together. Among the four fundamental forces of nature, the *strong force* has the greatest strength. Its strength is around 10^2

times that of the electromagnetic force, some 10^6 times as great as that of the weak force, and about 10^{39} times that of gravitation. However, it is a short-range force unlike gravitation, the weakest of the four fundamental forces.

Among the first physicists, who worked on the idea of the *strong force*, was Hideki Yukawa (1907–1981). In 1935 he published His theory of *mesons*, which was published in 1935, explained the interaction between protons and neutrons. It was also a major influence on research into elementary particles. *Quantum chromodynamics* (*QCD*) describes the strong interactions. It deals with a property of matter called *color* (it is analogous to the electric charge and has nothing to do with color in the visual sense of the word).

The electromagnetic force and the weak force were unified by Abdus Salam, Steven Weinberg and Sheldon Glashow. Prediction of the existence of the Higgs boson was an indirect aspect of the *electroweak theory* developed by them. *Chiral symmetry*, which Salam had introduced in the theory of neutrinos, first postulated by Wolfgang Pauli in 1930s, played a crucial role in subsequent development of the *electroweak theory*.

Physicists have come across a large number of fundamental particles over their course of research. Particles are also thought to act as *messengers* for the fundamental forces of nature. In current understanding, particles are excitations of quantum fields and interact following their dynamics. The *string theory* takes an alternative turn. Unlike particle physics, the *string theory* does not hold particles to be fundamental. It considers the different types of the observed elementary particles of the quantum world to be different quantum states of tiny strings. The *string theory* was first used in an attempt to explain the *strong force*. However, it was replaced by *quantum chromodynamics* that does the job way better than the *string theory*. The next field that the *string theory* was attempted to be used was to find a theory for *quantum gravity* (the force of gravity according to the principles of the *quantum theory*).

An extended version of the *string theory*, called the *M-theory*, was formulated later on and projected as the *Theory of Everything* in physics by its proponents. The *M-theory* identifies 11 dimensions of spacetime. These 11 dimensions include 7 higher dimensions apart from the 4 common dimensions (3 of space and 1 of time). Edward Witten proposed the *M-theory* at the annual conference of string theorists at the University of Southern California in 1995. He, however, drew from the work of a number of prominent string theorists for this. The list of these string theorists is long. Just to name the most important ones, we must name Christopher Michael Hull, Ashoke Sen, Paul Kingsley Townsend, Michael James Duff and Yoichiro Nambu, Michael Boris Green, Gabriele Veneziano, Leonard Susskind, Joël Scherk, John Henry Schwarz.

Ashoke Sen, in his paper on the *strong-weak coupling duality* or *S-duality*, showed for the first time the equivalence of all the various existing string theories of the time. He also has notable contributions to the study of *D-branes* (a class of extended objects upon which open strings can end with fixed boundary condition). The *Sen conjecture* about *D-branes* states that the tachyons carried by open strings attached to *D-branes* reflect the instability of the *D-branes* with respect to their complete annihilation. A *brane* may be thought as the projection of a point like particle in multiple dimensions. *D-branes* are a very important concept in the *M-theory*, which follow the *Dirichlet boundary condition* or fixed boundary condition. A boundary condition, in the mathematics differential equations, refers to the restraints that the values generated must obey.

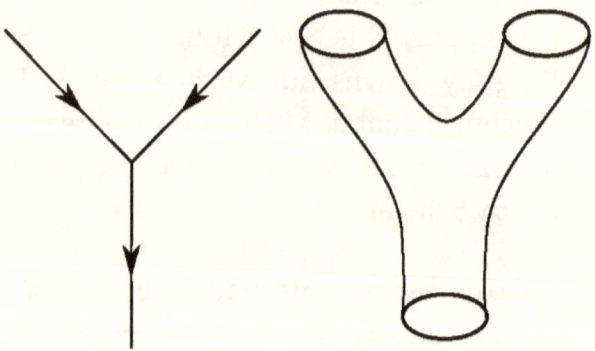

Fig. 7: World lines and world sheet - for the *string theory*.

Many physicists have expressed their skepticism about the string theory and its developments. They even doubt the testability of the theory. The extremely small predicted upper limit for the size of the strings that constitute particles is one of the greatest hurdles in devising experiments to prove their existence. Strings are typically believed to have sizes close to the *Planck length* ($1.616199(97) \times 10^{-35}$ metres). Other difficulties include the Lack of uniqueness of predictions due to the large number of possible solutions.

Critics of this theory include physicists like Richard Feynman, Lawrence Krauss, Sheldon Lee Glashow, Peter Woit, Philip Warren Anderson, Roger Penrose, Carlo Rovelli, etc. Roger Penrose has this to say about the theory:

I don't like that they're not calculating anything. I don't like that they don't check their ideas. I don't like that for anything that disagrees with an experiment, they cook up an explanation--a fix-up to say, "Well, it might be true." For example, the theory requires ten dimensions. Well, maybe there's a way of wrapping up six of the dimensions. Yes, that's all possible mathematically, but why not seven? When they write their equation, the equation should decide how many of these things get wrapped up, not the desire to agree with experiment. In other words, there's no reason whatsoever in superstring theory that it isn't eight out of the ten dimensions that get wrapped up and that the result is only two dimensions, which would be completely in disagreement with experience. So the fact that it might disagree with experience is very tenuous, it doesn't produce anything; it has to be excused most of the time. It doesn't look right.

2

The Standard Model of Physics

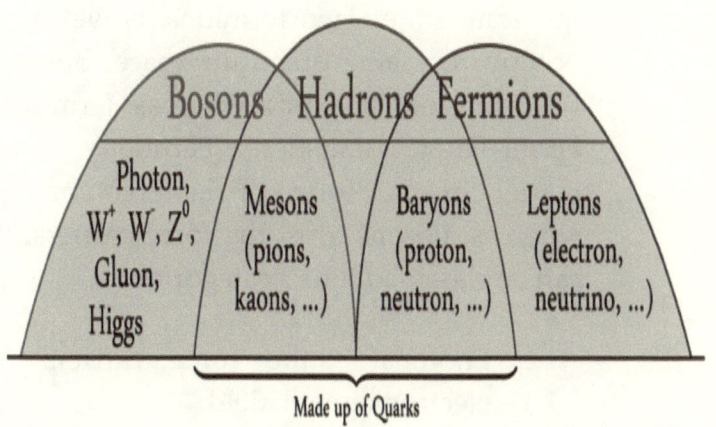

Fig. 8: Particle classification: bosons, hadrons and fermions.

Fig. 8 shows the 3 basic groups of particles in the Standard Model of particle physics. There are more than 60 elementary particles in this model that describe the behavior of the electromagnetic force, the weak force and the strong force. Actually, the particles are classified into two main heads – the bosons (named after Satyendra Nath Bose) and the fermions named after Enrico Fermi). One of the most important characteristic of bosons is that their statistics do not restrict the number of particles that can occupy the same quantum state. Two fermions, however, cannot occupy the same quantum space. Bosons obey Bose–Einstein statistics, whereas, fermions obey Fermi–Dirac statistics. Fermions constitute matter in the form of leptons and quarks, whereas, bosons form the force carriers. Bosons can be classified into 3 categories:

1. Photons – the force carriers of the electromagnetic field.
2. W and Z bosons – the force carriers of the weak force.
3. Gluons – the fundamental force carriers of the strong force and binds matter together.

The Higgs boson that is believed to be the particle that imparts mass to other particles. On 4 July 2012, the CERN Large Hadron Collider officially announced that the Higgs boson was observed. However, further tests may be required to confirm this discovery.

In the Standard Model, the fermions are considered to be of two types: the

1. Quarks – a fundamental constituent of matter (by forming hadrons).
2. Leptons – particles with spin-½ that are unaffected by strong interactions.

There are 24 leptons in all that include six quarks: the *up quark, down quark, strange quark, charmed quark, bottom quark,* and *top quark*; and six leptons (*electron, electron neutrino, muon, muon neutrino, tau particle, tau neutrino*), each with a corresponding *antiparticle*.

One of the most important properties of leptons is the electric charge that they carry. The leptons have no intrinsic mass and can only acquire mass through interactions with the Higgs field.

The strong force holds quarks together in order to form the hadrons. Basically, the hadrons into two broad categories:

1. Baryons – made up of 3 quarks (protons and neutrons are baryons).
2. Mesons – made up of 1 quark and 1 antiquark (pions, which Pions also play a role in holding atomic nuclei together).

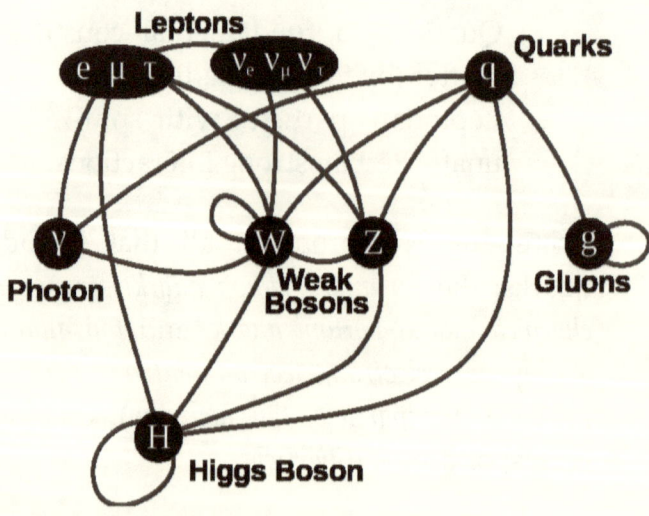

Fig. 9: Interactions of elementary particles, described by the Standard Model.

Neutrinos (meaning the *little neutral one* in Italian) are a class of particles that carry no charge and zero (actually, very little) mass. Neutrinos are affected only by the weak force (very short range force) and gravity (which is relatively weak on the subatomic scale). Neutrinos, therefore, pass through normal matter mostly unimpeded. As such, they cannot be detected directly. Neutrino detectors are often built underground in order to isolate the detector from cosmic rays and other background radiation that might hinder its detection. There are three types, or *flavors*, of neutrinos: electron neutrinos, muon neutrinos and tau neutrinos. Neutrinos are being created all the time, in huge numbers, in the Sun itself. Radioactive processes cause neutrinos to be generated. Millions of them may be passing through your body at this very instant. They are passing through you practically unhindered; they are so small as compared to other particles.

The messenger particles of the strong force are the gluons. These particles thus act as the *glue* that binds the quarks together and form the hadrons, which in turn form most of the matter that we see around us. Gluons carry the *color charge* of the strong interaction.

The muon is an elementary particle, which is similar to an electron, with greater mass than an electron. These particles are very short-lived, with a mean lifetime of 2.2 μs (microseconds).

The W and Z bosons (or the weak bosons) are the elementary particles that mediate the weak interaction. The W boson can have one unit of positive charge (W^+) or one unit of negative charge (W^-). The Z boson is electrically neutral, however.

The Standard Model is a very successful theory of fundamental interaction. However, it is not free from its own set of assumptions. As such, there are various shortcomings too in this theory. Just to give a few examples, it does not account for an expanding universe (as described by Hubble's law), nor does it account for gravity (as described by general relativity). It also does not correctly account for neutrino oscillations (and their non-zero masses). *Dark matter* and *dark energy* are two other matters that the Standard

Model remains silent about. The model requires 19 numerical constants at present, whose values are unrelated and arbitrary. Many physicists consider this an ugly. In order to account for the Higgs field and its behavior, the Standard Model would need to be twitched. Self-consistency of the Standard Model is also something that is in question. There are other very fundamental questions that remain open inside the Standard Model. The most fundamental of them all is, perhaps, why the Standard Model exists and what gives rise to it. Another question that remains unanswered is why the universe seems to have more matter than antimatter. Whys the particles in the Standard Model have their respective masses and their other respective properties is also not answered satisfactorily by the theory.

3

A Few Open Problems Today

Dark Matter:

The observed mass of galaxies, including the Milky Way, is far less than what should be there in order to account for the gravitational effects that we see. It was first postulated by Jan Oort in 1932 to account for the orbital velocities of stars in the Milky Way and by Fritz Zwicky in 1933 to account for evidence of *missing mass* in the orbital velocities of galaxies in clusters. We cannot see dark matter and it is believed that it does not emit or absorb (or even interact with) electromagnetic radiation. In 1975, at a meeting of the American Astronomical Society, Vera Rubin and Kent Ford announced their discovery that most stars in spiral galaxies orbit at roughly the same speed, implying that the mass densities of the galaxies were uniform well beyond the regions containing most of the stars (the galactic bulge). The Galaxies rotate with such speeds that the gravity that can be accounted for by visible matter should prove much less than what is required to hold them together and to prevent them from disintegrating. The same phenomenon is observed in galaxy clusters. There too the constituent galaxies should have extra mass, than what is accounted for by visible matter, to explain why the clusters hold together and have not been torn apart long ago.

Dark Energy:

Observations indicate that the universe is expanding at an accelerating rate. Something out there, which we cannot see, seems to be pushing away the galaxies way faster than what can be accounted for. Dark energy is the alleged culprit. This dark energy seems to be evenly distributed throughout the universe and it does not interact through any of the fundamental forces of nature other than gravity. It seems to be not very dense, however (roughly 10^{-27} kg/m^3). This very low density of dark energy makes its direct detection very much unlikely with the facilities that we have today.

It is estimated that ordinary matter makes up only up to 5 percent of the universe. The rest is made of dark matter and dark energy (dark matter – roughly 27 percent and dark energy – roughly 68 percent) **Fig. 10**.

Fate of the Universe:

What is indicated by all observations is that the universe is expanding. The rate of this expansion seems to increase as we look further and further into the universe. Theorists reason that if the universe is expanding, then there must have been a time, when the whole universe was contained inside a single point. Then time started and with that the universe with the Big Bang. Georges Lemaître first proposed the Big Bang theory in 1927. Arno Penzias and Robert Wilson discovered the *cosmic microwave background radiation*, which is almost exactly the same in all directions, in 1964, favoring the Big Bang model.

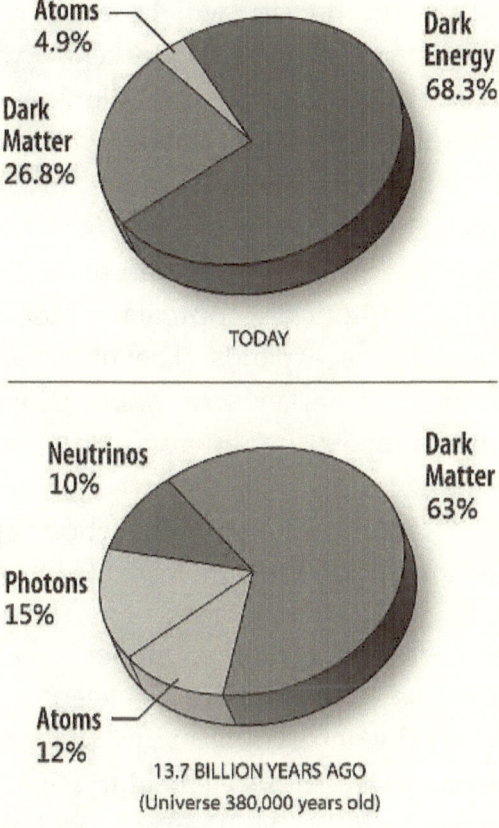

Fig. 10: Contents of the Universe, courtesy NASA.

Zero Postulation and the Laws of Nature
Subhajit Ganguly

There are many theories regarding the final fate of the universe. Some cosmologists favor the *Big Crunch* model, which states that after the universe has expanded to some maximum value, its average density will be enough to stop its expansion. Then the universe would start contracting (into a dimensionless singularity according to some). However, the Big Crunch model does not find much favor currently among cosmologists. The *Big Freeze* model suggests that the temperature of the universe will asymptotically approaches absolute zero, as the universe expands. The universe will grow darker and darker. *Heat death*, which is a state when temperature differences may no longer be exploited to perform work, will result. In such a state, there will be no information processing or life anywhere in the universe.

The *Big Rip* is another model, which suggests that as the universe keeps expanding, eventually, every galaxy in the universe will disintegrate into elementary particles and radiation. In the Multiverse Model, the universe is not one but many. These universes cannot contact each other. While, one universe reaches *heat death*, there are always other universes that have not yet reached it. Thus, in this model, the whole multiverse never ends.

Cosmic Inflation:

This suggests that in the early universe, space inflated between 10^{-36} seconds after the Big Bang to sometime between 10^{-33} and 10^{-32} seconds, at a rate much faster than the speed of light. Cosmic inflation is meant to explain the origin of the large-scale structure of the cosmos. We are, however, not sure if this theory is correct. If this theory is true, we are not sure how exactly this happened and what were the reasons behind it happening.

Horizon Problem:

As we look at great distances into the universe, it seems strangely homogenous. This is unexplained by the generally accepted *Big Bang* model. Cosmic inflation is regarded as the plausible explanation for this observation. As cosmic inflation itself is not something that we are sure of, variable speed of light is also given as a possible explanation by some. But then again, a variable speed of light would mean rewriting most of the physics that we know today.

Gravitational Waves:

General relativity predicts the existence of gravitational waves, which are ripples in the curvature of spacetime. We have found many indirect evidences of these waves. However, these waves have never been detected directly. The detection of gravitational waves is still an open question in physics.

Baryon Asymmetry:

The universe, as we see it, contains more (baryonic) matter and anti (baryonic) matter. The *Big Bang* should have produced equal amounts of matter and antimatter. We do not know the reason behind this baryon asymmetry.

4

Deduction and Emergence

Reductionism, in the theorizing process, attempts to break down complex ideas into simpler component concepts and holds that the whole is a sum total of its parts (see **Fig. 11)**. In the reductionist method, the phenomena that can be explained completely in terms of relations between other more fundamental phenomena are called *epiphenomena*. Reductionism is able to give more accurate predictions of the world, when the processes involved are linear. It is not that efficient in predicting the outcomes of nonlinear phenomena, however. The output of a nonlinear system is not directly proportional to the input, while the output of a linear system is directly proportional to the input. Nonlinear systems are of great interest to us today as most systems are inherently nonlinear in nature. Some aspects of the behavior of a nonlinear system appear commonly to be chaotic. Predicting the weather is one example, where such chaotic aspects are involved. Chaos theory studies the behavior of dynamical systems that are highly sensitive to initial conditions.

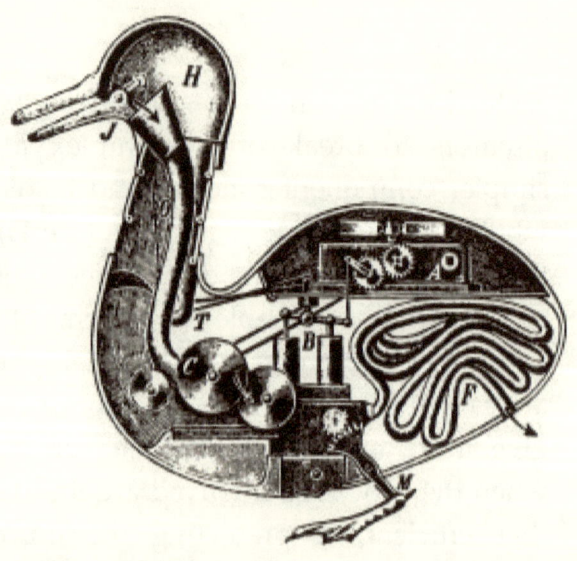

Fig. 11: Descartes held that non-human animals could be reductively explained as automata — *De homine*, 1662.

In a chaotic system, a very small change in the initial conditions can result in large differences in a later state. This is popularly known as the *butterfly effect* (**Fig. 12**), which derives its name from the theoretical probability of the formation of a hurricane, due to a butterfly flapping its wings few weeks earlier. A more commonplace demonstration may be, water may run down the slopes of a hill, depending on slight differences in its position, into any part of the surrounding valley.

Emergence is the way complex systems and patterns arise out of a multiplicity of relatively simple interactions. This is central to the study of complex as well as chaotic systems. The simple components of the environment may form more complex emergent properties, as a whole. Relating to the closest we can get, the human perception of the world, we can say that what we perceive about the world is way greater than the signals that are fed to our bodies through our senses. When you look at a distant star, your perception of the star is way greater than the light that reaches your eyes from it.

Fig. 12: A demonstration of chaos, using trajectories and the Lorenz attractor (named after Edward Lorenz).

Fig. 13: Snowflakes (by Wilson Bentley) forming complex symmetrical patterns is an example of emergence in a physical system.

Fig. 13, showing complex patterns being formed in snowflakes, is an illustration of emergent phenomenon. Emergent behaviors can occur due to intricate casual relations across different scales and feedback. Thus, complex patterns can emerge due to the interconnectivity of the components of the system. In complex systems, as the number of components under consideration increase, so does exponentially the number of interactions between them.

Thus, it becomes increasingly difficult for us to predict emergent phenomena, as there is an increase in the number of properties that we have to consider for doing that. However, it should also be kept in mind than merely a large number of insignificant components, coming together and constituting a given system, is not enough to ensure that we have emergent properties. The *noise* that is produced may sometimes subdue the emergent properties too.

One example of emergent phenomenon from the quantum theory is the fact that instead of a single definite momentum or position, the wavefunction describes a more fundamental state of matter. Mass is sometimes considered to be an emergent property, arising due to the interaction of particles with the *Higgs field*. The laws of physics themselves, as we know them, seem to be emerging from more fundamental laws of nature. All the rules of nature seem to emerge as a consequence of some *most fundamental rule*.

5

Zero Postulation and the Theory of Abstraction

Over some years now, a large part of the energies of the scientific community has been employed solely for finding a theory that will fit in all known happenings of the physical world. Various groups of scientists have tried to attack the problem from different ends. Some of these theories have been partly successful in explaining the known physical world. However none of these theories have been without shortcomings. Be it the much lauded String Theory or the Quantum Gravity postulation or any other such attempts towards arriving at a Theory of Everything, none have been proved to be foolproof. To say the least, nobody can deny that there is room for much improvement before we can even start thinking truly towards such a theory that would describe the known world satisfactorily and provide for a single basis of understanding the four forces in nature.

On top of that, we have the newly emerging problems of 'Dark Energy', 'Dark Matter' and the like. These realms are yet to be accepted by the scientific community officially, but nonetheless, they are most definitely at least a few parts of mysteries that remain unexplained. A good and effective Theory of Everything must aim towards explaining such mysteries too. Sadly, we have no theory as yet that fulfills these criteria.

From the dawn of civilization, human beings have tried to find out order in the chaotic world surrounding them. It has however never been easy to find a solution to explain a given system while being a part of that system. The best bet is to find out the most fundamental components within the system and building a theory round these. In other words, a theory that is able to describe the world in totality has to keep the number of basic postulates it depends upon to zero or near zero. Reductionism hits a dead end in this regard.

On the other hand, abstraction as the starting point of building up a theory may be seen to be of fitting use. It would be much more than a new way of tackling the problem. Even abstract postulates do away with the shackles that bind our theories into the system and bar them from being total descriptions of the system. The abstraction we are talking about here may be defined as, "Postulation of non-postulation" or, in other words, "A system of postulation that gives equal weights to all possible solutions inside the system and favors none of such solutions over others."

Abstraction automatically gives rise to optimized solutions within the universal set of all possible solutions, as has been shown in this book. It is these optimized solutions that make up and drive the non-abstract parts of the world, while the non-optimized solutions remain 'hidden' from the material world, inside the abstract world.

Starting from a basis of no postulation, we build our theory. As we go on piling up possibilities, we come to a similar basis for understanding the four non-contact forces of nature known till date. The difference in ranges of these forces is explained from this basis in this book. Zero postulation or abstraction as the basis of theory synthesis allows us to explore even imaginary and chaotic non-favored solutions as possibilities. With no postulation as the fundamental basis, we are thus able to pile up postulated results or favored results, but not the other way round. We keep describing such implications of abstraction in this book. We deal with the abstraction of observable parameters involved in a given system (quantum, relativistic, chaotic, and non-chaotic) and formulate a similar basis of understanding them.

Scaling of observable parameters in adequate ways is shown to unite the understanding of worlds of the great vastness of the universe and the minuteness of the sub-atomic realm. Finally, the mysteries involving 'dark energy' and 'dark matter' are uncovered using such an approach.

Considering transport or tendency of transport of physical entities from an initial to a final point, we come to a similar basis of understanding of various physical phenomena. The trajectory-behavior of such transport represents the effect or field of influence. This way, we may explain cluster-formation in the universe, an expanding universe, etc. This may also lead to a similar basis for understanding the four non-contact forces of nature. Also, for different ranges of acceleration in the field formed in space-time, we have different properties of matter interacting. This may explain the difference in ranges of the various forces.

All phenomena we see around us are but a physical or imaginary transport of various quantities starting from the flow of gases, to formation of fields of gravity, magnetism, etc., photons moving n space-time, right to the formation of cyclones and hurricanes, occurrence of events may be considered to be simply the transaction of physical or abstract entities between two points—one 'initial' and the other 'final'. The transfer of these various quantities, however, may not always be considered simple, and the 'route' of the transfer may form a complex relation with what may be called an environment. These transactions, however complex, must be obeying some physical laws, though, and all of them must be measurable for all physical considerations.

There must be a same set of physical and mathematical laws governing all such transactions in every frame of reference. Consequently, we may consider the same set of a single mathematical construct, a set of equations that can safely describe all these phenomena satisfactorily. That concerned set of equations may take different forms, however, in different sets of events, but there must be a singular basis for all these. In a nutshell, we must be able to conjure up a mathematical idea that can explain such transfers of all known physical entities.

Developments in the mathematics of 'Chaos Theory' have pointed towards statistical analysis not to be always 'deterministic'. Meaning which, we cannot always predict a single solution to a given transfer in consideration, if it tends into the chaotic region of prediction. This however does not restrict us in finding mathematical tools for describing all possible routes that can be taken up by a given system in reaching a final point from an initial one. Once we take into account all such possible routes of transfer or transactions, the safety of predictions increases to a maximum.

Such a transfer of any physical entity will continue until and unless the difference in concentrations concerned, i.e., λ becomes zero. Considering the example of heat transfer from a body at a higher temperature to one at a lower temperature, heat will continue to flow towards the colder body, until and unless the difference in concentrations, i.e., the difference in temperatures of the hotter body and the colder one becomes zero. Similarly, a given body will keep moving with uniform velocity in a straight line, until and unless there is an acceleration or retardation (support or resistance, respectively, as the case might be).

In general, the following may be postulated:

1. Any given physical quantity flows from a region of higher concentration of it to a region of lower concentration of it, until the concentrations in both the regions are the same.
2. Support (S) towards the concerned flow being zero in a given direction, the rate of flow $\left(\frac{dF}{dT}\right)$ is zero, as there is practically no flow in that given direction.
3. Resistance (R) against the flow being zero, time taken for the completion of the concerned flow is zero. Thus, if no resistance is present against the occurrence of a given event, it occurs practically instantaneously.
4. Considering any given region as a collection of a finite or an infinite number of initial and final points, we can say that any given entity within the region tends to flow equally in all directions. However, support and resistance towards and against a flow in any given direction will decide the routes of such a transport.

Instability in a System:

The logical approach that has been undertaken in studying transfer of a given entity between points seems to have led us to what may be called super-stability. As the given physical entity tends to flow equally in all directions, it seems that the universe, as we know it, should have been highly stable, with almost equal quantities of all physical quantities spread equally everywhere. Both matter and energy should have been in equal or almost equal proportions everywhere. But this is not so. The very logic concerned may be made use of to explain this seeming anomaly.

Let us consider the example of a three-point isolated system. Let the points be 'A', 'B' and 'C'. Let A and B be material points, whereas, C be situated anywhere on the straight line joining A and B. The material parts of both A and B tends to move in all possible directions. These possible directions include the directions towards each other. Thus, at point C, for obvious reasons, an additional effect will be felt due to the tendency of material to flow from A to B and from B to A, as compared to all other directions.

The points A, B and C being considered parts of an isolated system and all three points being assumed fundamentally similar (with the only difference that A and B contain material, while C is empty), the factors R and S must be equal.

Making use of a few useful concepts, the effect of material-points on their surroundings can be studied. Also, it is important to note that the presence of material-points ensures trajectories of all possible routes of the transfer of entities and these may be even chaotic in nature. The geometry around such points must therefore depend upon the probabilities concerned of the given trajectories revisiting the attractor-point. Transport of any physical entity (or even if it tends to be transported) will therefore affect the geometry of the surroundings in a similar way. Thus, we are bound to have fundamentally similar trajectories in all cases of transfer of entities. This being true, an actual transport or a 'tendency' of transport of any physical entity and its effect may fundamentally be predicted in a single similar fashion. By selecting the initial and the final points or regions concerned effectively, in any given scenario, we must be able to describe the behavior of the transport to the desired level of accuracy, by studying the trajectories of it. Choosing appropriate factors too should be necessarily sufficient in taking care of the desired level of accuracy in calculations.

The tendency of the material-points to flow towards each other, as discussed, must be sufficient to explain the absence of super-stability in the universe. Also, the non-material or empty points in the vicinity of the material-points, as discussed earlier, tend to behave just as material-points, thereby causing a change in the whole vicinity or environment concerned. This predicts the presence of 'fields' in the environment of material and non-material points, further nullifying the super-stability of the universe, explaining why we see the world the way we see it.

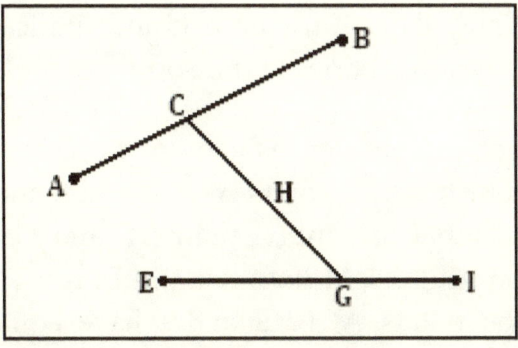

Fig. 14: **Four material-points A, B, E and I, and two non-material points C and G in their respective vicinities**

Using similar concepts, the probabilistic plots of all possible routes of transport are to be found, for a given scenario. These plots, in turn, yield the description of the effect or field of the concerned transport or transaction. Depending upon the transport being considered, we may have the respective sort of field, viz., electromagnetic, gravitational, etc. These fields, however, being fundamentally similar probability plots of all possible routes of transactions of physical or imaginary entities must have a similar basis for comprehension.

Thus, there must be a grand unification of all physical theories of transport.

Let us consider, as shown in the figure above, four material-points A, B, E and I, and two non-material or 'empty' points C and G, anywhere on the straight-lines AB and EI respectively. Let the points A, B, E and I have equal factorial conditions of material transport (equal to F) difference in concentrations of material (equal to λ) and concerned time (equal to T). Also, let AB and EI be having equal straight-line distances (equal to D).

The points C and G, and as a cosequence the point H too, inspite of being considered empty points, are seen to behave just as material-points in the vicinity of material transactions. Effects of such a transaction affect non-material points and these non-material points in return can thus affect other material or non-material points. Thus, the points themselves that we consider, irrespective of being empty or containing material, behave as if being in a same kind of transport themselves. Therefore, we are to deal with not only real material transport, but transport of abstract entities too. Not only must a physical entity tend to move towards all points in its vicinity, the points that we consider, themselves must tend to be moving in all possible directions as we can see. The points themselves will have to be studied therefore by their trajectory behavior.

Using desired scaling-lengths, the necessary factorial conditions for a given transport may be estimated to the necessary level of accuracy, for performing calculations. Each of the quantities that affect transactions may have a given number of dimensions for a given level of acccuracy in predicting the routes of transaction. A desired level of accuracy for a given set of measurements can be achieved by knowing exactly the necessary parameters for the observations.

At a lag of 0, the co-ordinates of each plotted point are equal, auto-correlation coefficient being maximum. Auto-correlation decreases with increase in lag. By adequate accuracy in observations for a desired set of measurements, a necessary level of accuracy may thus be achieved.

Spillage in all possible directions may be prevented, for a given scenario, by the factorial conditions *R* or *S* for each of these directions.

An expanding Universe:

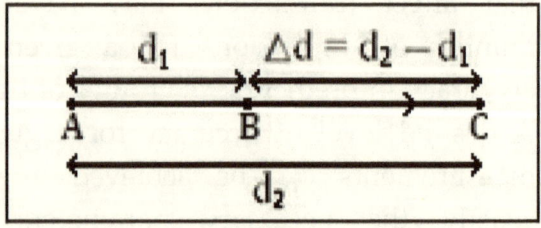

Fig. 15: Relative motion of a point from B to C, in vicinity of point A

Let us consider a point of reference A. Let another point move from point B to point C, with respect to point A. Let B and C be at distances d_1 and d_2 respectively from point A. Considering $S = R$ for the flow, and λ to be a constant (considering the movement of the concerned point from one 'empty' place to another),

$$\frac{F}{T} \propto \frac{1}{D}$$

Thus, as the flow takes place from B towards C, as the distance between B and C ($\Delta d = d_2 - d_1$) decreases, the velocity of the flow $\left(\frac{F}{T}\right)$ increases. Thus, an observer at A will see that the velocity of the moving point goes on increasing as it keeps moving away from A. This is in consistency with Hubble's law ($v = Hr$), for an expanding universe.

As the point B keeps moving away from A towards point C, the effect of point A on point B goes on decreasing. The effect of the field around the reference point seems to lessen thus with distance, getting the points further from it to spread away with increasing velocity. The distance between the reference point and the moving point being sufficiently greater than the 'size' of the points concerned, this spreading away from the reference point assumes considerable importance for obvious reasons.

The Four Non-Contact Forces:

The four non-contact forces, viz., the electromagnetic, the weak, the strong and the gravitational forces, being considered the effects of transport of respective entities, may be treated in a similar fashion. The equations would yield the probabilistic plots or fields of all routes of actual transport or tendency of transport in each case. When the distance between the concerned initial and final points of transport become sufficiently large, however, than the 'size' of the points, the trajectories tend to become chaotic, thereby increasing the uncertainty of predictions. This uncertainty, in turn, depends upon the Lyapunov exponents (v). Moreover, there is a stretching or shrinking of a given direction according to the factor e^{vt}, according as v being positive or negative in that direction.

The prediction time increases only logarithmically with the precision of the initial measurement. Thus, in such chaotic states, where the size of the concerned points becomes sufficiently small or large as compared to the concerned distance inbetween, only short-term predictions are possible.

Let us consider the introduction of a given physical property, viz., mass, charge, colour, etc., in this region. The vicinity will be affected by this introduction. If the size of the concerned source of disturbance is very small or very large as compared to its vicinity, chaotic states prevail, and no long term prediction will be possible.

The change in the space-time configuration will be caused as all the points in the vicinity of the source of the concerned physical property is affected by its field, as stated earlier.

An acceleration will be felt by the concerned physical property in the vicinity of the source. As such, there will be a force for each of these properties concerned, viz., gravitational force for mass, electromagnetic for charge, colour-force for colour, etc. The value of time in the space-time is different for each of such interactions.

Therefore, the strength of the interacting force for each property is different too.

The strength of this force, as stated earlier, will be different for each concerned physical property. However, there must be some similar basis of interaction with each of these strengths of forces. The trajectory behavior for each of these fields of force may be studied using the Lyapunov exponents.

Also, the quantity of force forming due to the interaction of a property A' in a given acceleration-field 'a' may be considered to be different. Interaction of different properties of matter, viz., mass, charge, colour, etc., with the same acceleration-field may yield forces of different ranges. Thus, it may be considered that at different ranges of acceleration different properties of matter will interact, although, these may do so in an equivalent manner. Therefore, there must be different ranges of forces that have influence over different properties. This is consistent with available data. The types of forces that are seen arise due to the interaction of respective properties with an equivalent acceleration-field.

Starting with studying actual transport or tendency to transport of entities, we have arrived at a conjecture, a theory that all physical phenomena may be explained through a single rule and thereby may be studied in a similar way. In each of these considerations, we take into account the initial and the final points of transport or flow. Choosing appropriate parameters for measurement and applying adequate observation, we can find the trajectories or the effects or fields of their behavior. For a given set of calculations, we will have to take care of the desired level of the accuracy for the intended set of results. Even if we conveniently omit a few parameters or information regarding such transport, we may be able to reach our desired level of accuracy. Reaching towards the complete set of parameters though, will ensure moving towards complete accuracy of predictions, even if the related trajectories become chaotic. The prediction-time in chaotic regions becomes less though. Placing the desired direction of flow of the given physical entity along a given direction, we may find the extent of flow that spills into any of the other remaining directions.

We have also explained cluster-formation, or the coming together of physical entities in the universe, and the stability of such systems. An expanding universe may also be explained in a similar way.

Finally, considering transport or tendency of transport of different physical entities, we may arrive at the unification of the four non-contact forces of nature. A force arises due to the interaction of a given physical property with an acceleration-field, created due to the same property or an equivalent one. Different physical entities interact with different ranges of the acceleration-field and this may give rise to different ranges of the various forces.

What seems to be lacking in the Standard Model and the General Relativity is a proper understanding of what energy really is. None of the two basic theories can satisfactorily tell us what stuff energy is really. While general relativity goes closer than the Standard Model in explaining energy, it only deals with the effect that energy has on the fabric of the spacetime, without giving a concrete definition of energy. The Theory of Abstraction, arising out of zero postulation, defines energy in these simple terms: *energy is all that there is*. It is the instability that arises out of that, which gives rise to all the known matter, the effects of energy and everything that we see around us. Thus, in a way, it is energy that *sees* energy with energy.

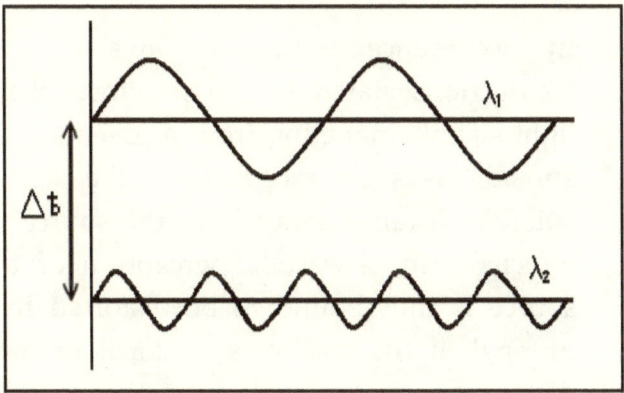

Fig. 16: Two waves of wavelengths λ_1 and λ_2 at a lateral distance $\Delta r b$ from each other. $\Delta r b$ is the minimum distance between the waves so that they are distinguishable.

Considering transport of light through space-time, following the laws of physical transactions, it may be said that there must be a spreading effect on it. Over suitable distances from a source of light, an observer's perception is bound to be affected due to this spreading. In the following chapter, these effects on the reception of a signal, due to the spreading of light are studied. Experimental set-ups are devised to verify the actual angles of spread with their theoretical values. An experiment regarding the minimum distance between two disturbances for them to be distinguishable is also carried out. The energy quantum is also studied in a new light.

In accordance with the laws of physical transactions stated in the previous chapter, a light-signal emanating from a given source will spread. As such, reception of the signal at a suitable distance away from the source will be affected. An observer's perception of a given source is thus bound to be distorted from the original, if the source is at considerably large distances away (example: reception of light from other galaxies). Over smaller distances between the source and the observer, though, the effects of the spread of the light-signal may not be considerable however, being of negligibly small dimensions.

ZERO POSTULATION AND THE LAWS OF NATURE
SUBHAJIT GANGULY

As an experimental set-up on an inter-galactic scale may not yet be possible, in order to examine the spread of light, we have to devise such a set-up which is possible and practical. This chapter also deals with such an experimental set-up, with which it is possible to verify the spread of different wavelengths of light. Theoretical values are tallied against the experimental ones. Further, the minimum distance of separation required for two waves to be distinguishable is studied through another experiment. Lastly, the energy-quantum is studied in accordance with the laws of physical transactions, with regards to the Theory of Abstraction.

Notwithstanding the effect of spreading of photon-gas, the photons themselves must tend to expand, according to the last chapter. Light with different wavelengths must have a proportional rate of spreading.

From previous arguments it is obvious that there must be a certain minimum distance between two given disturbances, so that they are distinguishable. This minimum lateral distance between the waves so that they can retain their individuality would depend on the difference in wavelengths ($\Delta\lambda$) for the two given waves.

Due to the introduction of an energy quantum, the vicinity gets 'stretched' from time T to T'.

Though the angles of spread of various wavelengths are quite small, yet making use of an intermediate wavelength between two given wavelengths in an experimental set-up allows us to detect such spreading. Experimental values tally with theoretical ones.

A minimum distance must separate two given wavelengths for us to distinguish them. This minimum distance depends upon the difference between the concerned wavelengths. Experiment to study this gives us the value of the constant of proportionality between the two.

The energy quantum is studied in the light of the Theory of Abstraction and the laws governing physical transactions. The Plank's constant (h) appears to be a very monotonously increasing function instead of being a constant in the strictest sense of the term.

Making use of the laws of physical transactions, we study symmetrical many-points systems. Relation of group-theory to physical transactions in such symmetrical systems is dealt with. Studying perturbations in the stability states in the attractor-maps for transactions, approximate values of the observables are to be predicted for such systems. Further, Abstraction Theory is typified with respect to studying the properties of irreducible representations, if any, inside a given such group.

A given point is influenced by its environment. On the other, it influences its concerned environment. A given point has some intrinsic properties. A group of such points forms a field of extrinsic properties. The field of extrinsic properties, in turn, may influence the intrinsic characters of each of the individual points.

A set of points with same or of a similar-set of properties may be considered to belong to a given same group. For any given system, there can be one or a number of stability-states or symmetries. Further, each of such symmetries may have perturbations, affecting the average value of observable quantities. Measures of such perturbations are a useful way of finding approximate functions for systems when we know the exact transaction-functions for similar systems.

Say a given dynamic system has a given set of symmetries or stability points. For all points having similar intrinsic properties within such a system, the probability densities of occurrence are equal and must remain unaltered, being all in a similar environment.

The sum of the squares of the dimensions of the irreducible representations of a group is equal to the order of the group, in accordance with group-theory.

Formation of Poles:

Each of the intrinsic properties contained inside a system will, for obvious reasons (in accordance with the Theory of Abstraction) transact in such a manner so as to be distributed as much as possible. This will give rise to two sets of transactions:

1. The transaction of the points themselves in every direction, including towards each other. This gives rise to an additional effect in the direction between two given points as compared to all the other directions. This is due to the fact that in the direction between the points, there is an effect due to both the points, while in all other directions (considering a two-point system) the effect is due to one point only as shown in fig.1

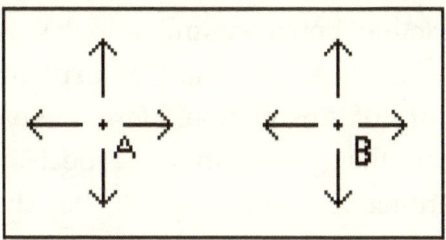

Fig. 17: Effect between points

This gives rise to an *attraction* between the points.

2. As a given intrinsic property transports in such a manner so as to be distributed as much as possible, the effect of 'repulsion' is generated. Let us consider two clusters of charges, either both positive or both negative. Not only the charge tends to be distributed in all directions, but also the type of charge. Presence of same type of charges will thus give rise to an effect of a field of repulsion in the vicinity of the clusters.

Thus we see that formation of poles and attraction between unlike poles and repulsion between like ones may be explained using the Theory of Abstraction. Here, we have examined only the qualitative aspects, while the quantitative aspects may be determined as detailed in my earlier chapter, in view of the laws of physical transactions.

Abstraction Theory is made use of to study symmetrical many-points systems. The intrinsic properties of each of the constituents of the system need to be treated in accordance. We may relate each of the concerned points inside the given system to various groups, as per their intrinsic properties. A group of such points i.e., a system having individual intrinsic properties inside it may be regarded as a point with its own intrinsic properties (a function of its the constituent intrinsic properties) when it is considered as a part of some larger system. This way, intrinsic properties of each of the smallest points may influence a system or even a group of systems. A group of such individual points may give rise to a field of extrinsic properties, affecting each of the individual points.

Considering the symmetries, or the stability states inside a given group or inside a given system, as being similar in the fact that they are all basically states of stability, the irreducible representations, if any, inside a given system may be studied. Perturbations regarding a given average observable value may be predicted in a similar way.

A particle in an isolated box will tend to move in all possible directions. A bias towards any given direction indicates an imbalance of support towards its movement in that given direction and resistance against it. Considering the movement of an energy quantum in a particular direction, this difference between the concerned support and the concerned resistance must be at least the same of the given energy quantum, in accordance to the Theory of Abstraction. For a given quantum-state ($h\upsilon$),

$$S \sim R = h\upsilon$$

where, S and R represent the support and resistance respectively.

This means that at least one half of a total energy-quantum gives it its direction while the other part gives it its magnitude. The direction part remains 'hidden' while only the magnitude part shows up as the value of the given quantum state. Considering the direction part however may reduce quantum-transport to classical transport.

From the Theory of Abstraction, we arrive at 'hidden' direction part of an energy-quantum. Quantum dynamics is seen to merge with classical dynamics if this hidden direction-part of the quantum-states are taken into consideration, as validated by practical analysis and data. Moreover, this hidden part of an actual energy-quantum may explain the dark-energy problem. As a support towards transport comprises the direction-part only, and as the resistance against motion is offered against the whole of an energy-quantum (direction-part + magnitude-part), this hidden energy may very well affect a gravitational field. On the other hand, if matter is considered to be condensed energy, and as such condensed energy quanta in some form of orientation in space time, there ought to be some self-same hidden mass as the magnitude-parts of the constituent energy quanta, thereby indicating a hidden dark-matter. Thus the anomaly of the existence of dark-matter and dark-energy being hidden from us may be reconciled with in accordance with the Theory of Abstraction.

The Abstraction Theory is applied in landscaping. A collection of objects may be made to be vast or meager depending upon the scale of observations. This idea may be developed to unite the worlds of the great vastness of the universe and the minuteness of the sub-atomic realm. Keeping constant a scaling ratio for both worlds, these may actually be converted into two self-same representatives with respect to scaling. The Laws of Physical Transactions are made use of to study Bose-Einstein condensation. As the packing density of concerned constituents increase to a certain critical value, there may be evolution of energy from the system.

Be it the large vastness of the universe or the delicate smallness of the sub-atomic world, by choosing a suitable constant scaling ratio for both, we may obtain their representations. These representations, following a certain constant scaling ratio, will be self-same. In the previous chapters, I have mentioned the chaotic behavior in the quantum world. Choosing suitable scaling ratios, we may turn the universe itself into such a chaotic quantum system, having its own necessary quantum states and trajectory behavior. In that case, the study of the universe reduces to the study of some sort of a quantum chaotic system. On the other hand, choosing some other necessary scaling ratios, the atomic and the sub-atomic realm may be extended to become the universe itself, complete with its own macroscopic trajectory behavior. Instead of formulating different ways of looking at worlds of different sizes, if we adjust the way of viewing i.e., the scaling ratios in such a fashion that the representations of the worlds merge, we will be looking at representative worlds of study which are practically self-same. The Laws of Physical Transactions formulated in previous chapters may then be applied in order to study such self-same representations of the worlds of various scales. Unification of the ways of studying at different ranges of scaling may thus be achieved by suitable landscaping (adjusting different scales to a suitable scaling ratio, in order to make all the scales of study similar in size). Further, a similar approach may be applied to study the Bose-Einstein Condensation. A certain critical packing density of the constituents of each world of a certain landscape must ensure a condensation of similar sort. The quantum states (or some similar states) of each such landscape will merge and give spikes for that critical scaling ratio in their respective representations.

The quantum chaotic behavior may be of interest to study if we are to learn about the universe as a whole. The astronomically large distances separating clusters in the universe supports a study of such sorts. Quantum chaotic behavior, on the other hand, will give rise to something similar to the Bose-Einstein Condensation at some critical packing density. The study of such condensation states too will be of interest here.

Looking at a large enough part of the universe, we may draw an analogy to a system of scattered particles, in motion or rest, relative to each other. These particles may or may not be similar to each other, if we look at a given locality. Our idea, however, is that we can always represent even the whole of the universe on a piece of paper of our desired size. We can very well do the same with localities of sub-atomic sizes.

We may represent both the worlds, viz., the microscopic and the macroscopic, within any desired standard size. Theoretically, we are only to diminish the snaps of the universe and magnify the snaps of the microscopic world in order to put both into representations of a definite scaling-size. Looking at such a representation of the macroscopic world (due to the large number of constituents and the large distances separating them involved) we will find it to be a complex mixture of various kinds of particles. On the other hand, looking at such a representation of the microscopic world, (due to the small distances separating the constituents) it will be like the actual universe itself, with various types of constituent parts involved. Such a representation of the microscopic and the macroscopic worlds will bring out hidden properties and behaviors of both worlds, as well as providing for a similar basis of studying them both.

A close enough representation of the constituents of the universe is affected by this loss in dissipation energy information inside its various clusters. This in turn will give rise to seemingly anomalous behavior inside the representation. The clusters will seem to move away from each other with greater velocities than anticipated values. On the other hand, the clusters themselves will seem to be bound with greater strengths than is anticipated. The existence of dark energy and dark matter that we feel may be linked to the loss.

Choosing a suitable scaling ratio we may represent the microscopic and the macroscopic worlds on the same scale, enabling us to study and compare their various hitherto hidden properties. A large enough packing density ensures formation of cluster points inside the representations. These cluster points will tend to be in their lowest energy dissipation states. The whole universe being considered as a cluster when its constituents are close enough, as in the moment of the Big Bang, it tends to be in its lowest energy state (theoretically a zero energy state). ΔI_ε, i.e., the difference in dissipation energy information which tends to infinity as the number of constituent points inside it tends to infinity, however establishes itself as Big Bang takes place. Yet, as the universe can expand further as its constituents move away from each other, with respect to a further expanded state it at any given present representation its clustered. Thus a hidden amount of energy dissipation information is present at any moment we look at the universe. This hidden energy dissipation information will make the clusters to move away from each other and the clusters themselves to be bound within themselves with greater hidden strengths than is anticipated.

Zero-postulation, as we can see, sits at the heart of abstraction. Null-postulation or zero-postulation favors no given result or a given set of results over all others. In that way, null postulation does not assume anything beforehand. What it does is to consider all possible results and derive the ultimate results from this exhaustive set of possible results. Each valid element inside the exhaustive set of results might interact in order to culminate into the 'real' results or happenings.

In not favoring solutions or sets of solutions, the principle of zero-postulation drives away any unwanted incompleteness from the description of the world. It is the interactions between the possible exhaustive set of solutions that creates the impression pointedness or directiveness in the universe, leading to the formation of clusters, as discussed earlier. These interactions may be chaotic in nature, giving rise to attractor points where the directiveness inside any given system asymptotically seem to approach. It is this directiveness, in turn, inside a given system or in the universe as a whole, that is the cause of all known phenomena. This directiveness of possibilities saves the universe from being exactly the same throughout, but makes it heterogeneously active, as we see it to be.

As zero-postulation considers the exhaustive set of all possible results, it would yield to be perfectly flexible to work with. That is to say, the scaling-ratio of observations may be adjusted as per requirements an intentions of the observer. This in turn is seen to unite observables at both microscopic and macroscopic levels through a similar basis of understanding. Even the 'non-real' abstract points themselves in the exhaustive set may be seen to interact and give rise to 'real' possibilities. Through analysis of all such possibilities, combined with analysis of all real elements inside the set is the description of the world in totality. The principle of null-postulation holds the key to such total description of the world, or of any given system, for that matter. Choosing suitable scaling-ratio and identifying all real and abstract parameters within a given system enables us to describe the system in all totality. Different scales of observations may have different sets of parameters with still different sets of interactions between themselves. These various levels of interactions between the parameters may give rise to different force-fields with their own respective sets of accelerations of interactions. However, all such fields of interactions, being fundamentally similar, have a similar basis of description.

In the description of the universe or any given system, there is always some 'hidden' information that does not show up while the rest of the available information is being taken into account. Different sets of such available information at various levels of scaling will have their respective sets of such hidden information. This may prevent a given description of the world, or a part of it, at a given scale, from being completely deterministic. Lack of complete determinism in directiveness inside a given system, or in the universe as a whole, gives rise to hidden mass and hidden energy, that may not seem to show up in a given set of observations, but is seen to affect the overall description, nonetheless. This 'indeterminism' in directiveness will make the clusters to move away from each other and the clusters themselves to be bound within themselves with greater hidden strengths than is anticipated. The hidden mass exerts a hidden gravity inside the cluster concerned. The hidden energy, on the other hand, will tend to draw the clusters away from each other.

For a large enough or a small enough scale of observations, this hidden amount of information concerned seems to be of paramount significance. The galaxies themselves are seen to be held together by some hidden masses, while they are found to move away from each other by some hidden amount of energy. On the other hand, in the subatomic world, particles are seen to be held together by very strong hidden forces, while they seem to be interacting with each other with some hidden amount of energies.

Zero Postulation and the Laws of Nature
Subhajit Ganguly

In both the cases, a certain amount of hidden uncertainty in available information is always found to be present. At a scale of observations, which is more 'akin' to our own 'normal' day-to-day scale of looking at the world, which lies in between such vastness or such minuteness, the hidden amount of information seems to be of less significance. As the scaling-ratio and the object it is supposed to measure become increasingly commensurate to each other, the amount of hidden information for that given scale of measurements diminishes in size.

Looking at a large enough part of the universe, we may draw an analogy to a system of scattered particles, in motion or rest, relative to each other. These particles may or may not be similar to each other, if we look at a given locality. Our idea, however, is that we can always represent even the whole of the universe on a piece of paper of our desired size. We can very well do the same with localities of sub-atomic sizes.

Zero Postulation and the Laws of Nature
Subhajit Ganguly

We may represent both the worlds, viz., the microscopic and the macroscopic, within any desired standard size. Theoretically, we are only to diminish the snaps of the universe and magnify the snaps of the microscopic world in order to put both into representations of a definite scaling-size. Looking at such a representation of the macroscopic world (due to the large number of constituents and the large distances separating them involved) we will find it to be a complex mixture of various kinds of particles. On the other hand, looking at such a representation of the microscopic world, (due to the small distances separating the constituents) it will be like the actual universe itself, with various types of constituent parts involved. Such a representation of the microscopic and the macroscopic worlds will bring out hidden properties and behaviors of both worlds, as well as providing for a similar basis of studying them both.

Zero Postulation and the Laws of Nature
Subhajit Ganguly

From the Theory of Abstraction, we arrive at 'hidden' direction part of an energy-quantum. Quantum dynamics is seen to merge with classical dynamics if this hidden direction-part of the quantum-states is taken into consideration, as validated by practical analysis and data. Moreover, this hidden part of an actual energy-quantum may explain the dark-energy problem. As a support towards transport comprises the direction-part only, and as the resistance against motion is offered against the whole of an energy-quantum (direction-part + magnitude-part), this hidden energy may very well affect a gravitational field.

Any change in scaling-ratio such that the constituents inside the system concerned have a large enough 'packing density', will yield the system, as a whole, to be in a 'condensed' state of being, at that given scaling-ratio. Again, a suitable change in the scaling-ratio so as to break away from the condensation state yields a different set of observed results, complete with its own subset of hidden results and hidden information regions. This may explain how a whole world of its own set of possibilities can just come into being from such a condensed state of existence as the scaling-ratio is suitably adjusted. At such a scaling-ratio, this world is expected to have its own set of heterogeneity, which seems to be just the opposite of the practically homogenous level of existence as in the condensed state. Coming into existence of such a world, as well as, any non-existence of such a world may be attributed to changing scaling-ratios. Only a suitable change in the landscaping and scaling-ratio is sufficient in describing any such existence or non-existence, as a whole. This, in turn, may be taken as the explanation of how the whole universe can begin from a homogenous 'nothing' at the time of the 'Big Bang'.

The relative periodicity of trajectories describing a given transaction seems to visit a given centre of possibilities or a given set of centers of possibilities. In the chaotic region, at least, any given event seems to have a periodic set of possibilities of happening. This implies the revisit of trajectories around one or a given set of attractor points. However, trajectories seem to stray from such points and not be approaching them exactly, for any given set of measurements. Adjusting respective scaling-ratios and choosing suitable parameters to a desired level of accuracy of predictions may ensure effective descriptions of a given system, a given set of systems, or the world as a whole.

ABOUT THE AUTHOR

Subhajit Ganguly is a physicist whose areas of expertise include the Theory of Abstraction. His contribution to the theory is noteworthy, to say the least. The other areas of science that he has made notable contributions to include astronomy, mathematics and the Chaos Theory. Zero-postulation is a new concept he has introduced to the theorizing process in sciences. He is an Advisor to the Figshare Open Science Platform and is an Ambassador for the Open Knowledge Foundation.

www.ingramcontent.com/pod-product-compliance
Lightning Source LLC
Chambersburg PA
CBHW030811180526
45163CB00003B/1228